Earth Science Week 2017: Earth and Human Activity

October 8–14, 2017
Highlights Report

Copyright ©2018 by
American Geosciences Institute.
ISBN: 978-1986798495

American Geosciences Institute
4220 King Street
Alexandria, VA 22302 U.S.A.
www.americangeosciences.org
703-379-2480

If you have comments concerning this report, please contact:
Ed Robeck, Ph.D.
Director, Department of Education and Outreach
Director, Center for Geoscience and Society
American Geosciences Institute
703-379-2480 x245
ecrobeck@americangeosciences.org

See Our News Coverage
Because of the large and increasing number of news clippings citing Earth Science Week activities and resources, the print edition of the print report no longer includes clippings. To view the hundreds of press releases and news items promoting awareness of Earth Science Week each year, please visit online at **www.earthsciweek.org/highlights**. Thank you for helping us in our efforts to conserve resources and protect the environment.

Table of Contents

Page	Section
2	Introduction
3	Key Partnerships and Efforts
10	Earth Science Week Toolkits
10	Web Resources
11	Newsletter
12	Contests
13	Earth Science Teacher Award
14	Focus Days
15	Special Events
16	AGI Promotions
16	Congressional Recognition
17	State Proclamations
17	Publicity and Media Coverage
18	External Evaluation of Earth Science Week 2017: Key Findings
19	Earth Science Week Sponsors
19	Earth Science Week Program Partners
20	Earth Science Week 2017 Events and Activities by State and Territory
28	International Events

Published and printed in the United States of America. All rights reserved. No part of this work may be reproduced or transmitted in any form or by any means, electronic or mechanical, recording, or any information storage and retrieval system without the expressed written consent of the publisher.

Front cover: 2017 ESW poster design for AGI by Angela Terry. Front and back cover background ©Shutterstock.com/Pakhnyushcha

Highlights Report: Earth Science Week 2017

Earth Science Week 2017 Photo Contest entry by finalist Andrew DePriest

Introduction

Held October 8-14, 2017, the 20th annual **Earth Science Week** celebrated the theme of **"Earth and Human Activity."** The 2017 event promoted awareness of what geoscience tells us about human interaction with the planet's natural systems and processes. Learning resources and activities engaged young people and others in exploring the important relationships that exist between human activity and the geosphere (earth), hydrosphere (water), atmosphere (air), and biosphere (life). The 2017 theme promoted public understanding and stewardship the planet, especially in terms of the ways people affect and are affected by these Earth systems.

"Human activity has a huge impact on Earth, and the possibilities open to humanity are, in turn, shaped by Earth's natural processes," says Geoff Camphire, AGI's Manager of Outreach. "The geosciences are essential for understanding how we can make the most of opportunities and manage challenges offered in areas such as energy, technology, climate change, the environment, natural disasters, industry, agriculture, recreation, and tourism."

AGI organizes Earth Science Week as a service to member societies, with generous help from partners that provide funding, donate materials, organize events, and publicize the event. Funding partners in 2017 included the U.S. Geological Survey (USGS); American Association of Petroleum Geologists (AAPG) Foundation; National Park Service; National Aeronautics and Space Administration (NASA); American Geophysical Union (AGU); Geological Society of America (GSA); Association of American State Geologists (AASG); Society for Mining, Metallurgy and Exploration (SME); AmericaView; Archaeological Institute of America (AIA); and Howard Hughes Medical Institute.

Earth Science Week **participation remained strong.** People in all 50 states and more than 16 countries participated in events and activities. The Earth Science Week website received over 633,000 page views in 2017. In addition, hundreds of people nationwide actively participated in the program's visual arts, video, essay, and photography contests.

Earth Science Week events ranged from educators teaching Earth science activities in their classrooms to open houses held at major USGS field stations. A detailed list of events can be found in the second half of this report. This list represents only events reported directly to AGI, so please notify Earth Science Week staff if your participation is not listed.

Additional events are highlighted on the Earth Science Week website (www.earthsciweek.org/highlights), which features press releases and other items by members of the geoscience community, as well as news media promoting Earth Science Week. Television and radio news programs covered Earth Science Week on local stations in several states. Each year, web, print, and broadcast media coverage of Earth Science Week reaches more than **50 million people.**

This report is designed to give an overview of the activities organized by AGI and other groups for Earth Science Week. We hope this information on 2017 events and publicity inspires you to develop your own activities next year. Please visit www.earthsciweek.org for event planning, materials, resources, and support. Contact Earth Science Week staff at info@earthsciweek.org for assistance in planning for Earth Science Week.

Summary of Activities

Earth Science Week 2017 Photo Contest entry by Bryant Nakasone

Key Partnerships and Efforts

Earth Science Week's success depends on the collaboration of key partners. In 2017, AGI pursued signature initiatives and forged partnerships with numerous organizations (listed alphabetically):

Educators seeking teaching resources and other support were directed by AGI to the **American Association of Petroleum Geologists** (AAPG) and the AAPG Foundation, both longtime supporters of Earth Science Week. In addition, Earth Science Week promoted awareness of AAPG's Distinguished Lecturer and Teacher of the Year programs. AAPG Student Chapters received kits. Program participants were encouraged to read and use "Visiting Geoscientists: An Outreach Guide for Geoscience Professionals," a handbook co-produced by AGI and AAPG's Youth Education Activities Committee. Program participants were encouraged to attend AAPG's Annual Convention and Exhibition and compete to win the AAPG Teacher of the Year award. AAPG's "Particle Size and Oil Production" activity was featured in the Earth Science Week 2017 activity calendar.

AGI supported the **Austin Earth Science Week Consortium**'s celebration in Texas in 2017 by donating posters and other educational materials. Hundreds of middle-school students attended a career fair and participated in activities.

The **American Geophysical Union** (AGU) continued its role as a supporting program partner in 2017 with the contribution of funds as well as expertise. Earth Science Week's 2017 activity calendar featured AGU's "Modeling Earthquake Waves" classroom activity. AGU's annual meetings, professional development workshops, programs for college students, print and electronic resources, GIFT workshops, AGU On-Demand, and "LEARN With AGU" video series were promoted through the Earth Science Week e-newsletter, website, and activity calendar. Earth Science Week staff also presented at AGU's Fall Meeting in San Francisco.

Earth Science Week 2017 promoted awareness of the **American Institute of Professional Geologists** (AIPG), an AGI member society that advocates for geologists and certifies their credentials. AIPG offers several PowerPoint presentations online for free download, presenting career information for young, newly graduated geoscientists. AIPG also provided a geologic-timescale bookmark for the educator kit.

Program participants were encouraged to go online to conduct a new activity called "Climate at a Glance: From Local to National Scale" that the **American Meteorological Society** (AMS) created in cooperation with NOAA's National Centers for Environmental Information. The activity introduces students to the NOAA Climate at a Glance website, which allows real-time analysis of monthly temperature and precipitation data nationwide.

Earth Science Week 2017 directed participants' attention to the annual celebration of Earth Observation Day, a STEM educational outreach event of **AmericaView** and its partners, which engage students and teachers in remote sensing as an exciting and powerful educational tool. AmericaView also provided a "Lands of Landsat" game poster dealing with satellite imagery for the educator kit.

The **Archaeological Institute of America** (AIA), a continuing Earth Science Week partner and supporter,

SUMMARY OF ACTIVITIES
• KEY PARTNERSHIPS AND EFFORTS

provided a classroom activity on "Humans and Water, Past to Present" for the program's 2017 activity calendar. In addition, the program promoted awareness of and participation in AIA's International Archaeology Day, which takes place annually on the final day of Earth Science Week.

Earth Science Week directed participants' attention to the **Association of American Geographers** (AAG), an AGI member society that offers an array of web resources for K-12 and college-level instruction. Materials support geographic approaches to Earth science education. For example, Geographic Advantage, an educational companion for the National Research Council's "Understanding the Changing Planet," shows students how geographers use maps to explore environmental change. AAG provided a fact sheet on its GeoMentor program and a postcard on its annual meeting for this year's educator kit.

The **Association of American State Geologists** partnered with AGI and the USGS to support Geologic Map Day during Earth Science Week 2017. State geologists nationwide made geologic maps of their states available on their websites for students to use in classroom activities on Geologic Map Day.

Encouraging educators to make use of the resources offered by the **Association of Environmental and Engineering Geologists** (AEG) to its members, Earth Science Week 2017 turned a spotlight on AEG's technical publications, section and chapter meetings, and special educator sessions at the AEG annual meeting.

Earth Science Week promoted awareness of a website of great value to educators, AGI's **Center for Geoscience & Society.** Largely through its Education Resources Network, the center enhances geoscience awareness across all sectors of society by generating new approaches to building geoscience knowledge, engaging the widest possible range of stakeholders, and creatively promoting existing and new resources and programs.

Program participants learned about three online videos by the **Center for Ocean Sciences Education Excellence** (COSEE) depicting dramatic changes in Alaska's marine ecosystems through interviews with scientists. The videos were produced by COSEE Alaska in cooperation with other geoscience organizations.

As promoted by Earth Science Week, the **Climate Literacy and Energy Awareness Network** (CLEAN) online portal stewards a major collection of climate and energy science educational resources and supports a community of professionals committed to improving climate and energy literacy. Key components include the CLEAN collection of climate and energy science resources, CLEAN guidance in teaching climate and energy science, and the CLEAN network of professionals committed to improving climate and energy literacy. For this year's Earth Science Week Toolkit, CLEAN provided a bookmark featuring a link to key resources.

Earth Science Week staff traveled to Houston and Denver to exhibit at Energy Day festivals hosted by program partner **Consumer Energy Alliance,** an advocacy organization that provides consumers with unbiased information on energy issues. AGI staff shared geoscience-based energy information with thousands of students, teachers, and other community members.

The **Critical Zone Observatories** provided an informational flyer on this National Science Foundation program, including an NGSS-aligned classroom activity on the "Gulf of Mexico Hypoxia (Dead) Zone," for the Earth Science Week Toolkit in 2017.

Earth Science Week participants were encouraged to celebrate **Earth Day** in April 2017 with classroom activities, experiments, and investigations exploring the science behind how the world works. Because Earth Science Week offers education materials, information, and tools throughout the year, school audiences were urged to make use of tools highlighting the theme of "Earth and Human Activity." In addition, program participants were invited to take part in a free live online event, hosted by the California Academy of Sciences in April, during which teachers and students were able to ask questions of "Academy Sustainability Scientists" studying these topics.

Science teachers were invited to take part in the third annual **Earth Educators' Rendezvous** in July 2017 at the University of New Mexico, Albuquerque. The event's combination of workshops, posters, talks, round-table discussions, and plenary presentations helped guide participants through a suite of interrelated challenges characteristic of Earth education in schools, colleges, and universities today.

For this year's Earth Science Week Toolkit, **EarthScope** provided an informational postcard on the personal stories of Earth scientists.

ExxonMobil, a longtime Earth Science Week partner, continued its support of the program. During summer 2017, ExxonMobil Exploration and AGI partnered to hold two five-day Earth Science/STEM Teacher Leadership Academies in Houston. Each academy — one for K-5 teachers and one for middle-school teachers

SUMMARY OF ACTIVITIES
• KEY PARTNERSHIPS AND EFFORTS

— provided educators with Earth science content, hands-on activities, resources and field experiences for them to use with their students in the classroom and with their colleagues in professional development settings.

Flyover Country provided an informational postcard on this "Place-Based Education" mobile app for geoscience for the Earth Science Week 2017 Toolkit.

Through Earth Science Week, participants learned about educational resources and programs of the **Geological Society of America** (GSA), a longtime program partner. Featured education and outreach programs included the GeoTeacher program, Teacher GeoVenture trips, the Distinguished Earth Science Teacher in Residence program, and GSA's GeoCorps America program. GSA also organized International EarthCache Day at the start of Earth Science Week 2017 and contributed as an active partner in the Geologic Map Day initiative. Earth Science Week's 2017 activity calendar included GSA's "How Dangerous Are Tsunamis?" activity. The educator kit also featured a GSA postcard on its popular EarthCache program.

The **Geological Society of London** (GSL), an AGI member society, provided two online resources for learning about key geoscience topics. Electronic map-based resources were the focus of GSL's Plate Tectonics page. In addition, a site was launched to accompany GSL's Rock Cycle online module.

Program participants were directed to **Geology.com,** a major Earth Science Week partner which provides a variety of geoscience materials including daily Earth science news, maps, an online dictionary of Earth science terms, and information on geoscience careers, as well as resources for teachers, including links to lesson plans from major Earth science organizations. Geology.com, in turn, covered Earth Science Week announcements, programs, and activities throughout the year.

Google, an ongoing Earth Science Week partner, contributed a "Earth: This Is Home" activity to the Earth Science Week 2017 activity calendar.

The **Howard Hughes Medical Institute** (HHMI), a continuing program partner, provided key materials for the Earth Science Week 2017 Toolkit. Prized components of this year's kit included a poster on "Discovering Keystone Species" and a "BiomeViewer" postcard featuring links to a new free app on biomes.

To help teachers and students delve into the science behind current events, Earth Science Week directed them to "Recent Earthquake Teachable Moments," a

Earth Science Week 2017 Visual Arts Contest entry by finalist Ian Lee

website of the **Incorporated Research Institutions for Seismology** (IRIS). These resources — including PowerPoint presentations, animations, and visualizations, as well as links to Spanish-language materials and USGS data — dealt with, for example, the magnitude-7.9 earthquake that shook Papua, New Guinea, on January 22, 2017. For the educator kit, IRIS also provided an activity sheet exploring the question "Can Humans Cause Earthquakes?"

As program participants learned, ethical behaviors and practices are of vital concern to geoscientists. The **International Association for Promoting Geoethics** (IAPG), an international associate of AGI, celebrated the first annual International Geoethics Day on the Wednesday of Earth Science Week 2017. To promote awareness of geoethics, IAPG offered documents such as The Geoethical Promise and The Cape Town Statement on Geoethics.

Earth Science Week 2017 promoted awareness of a new free app by the **Lamont-Doherty Earth Observatory,** called "Polar Explorer: Sea Level," which lets users explore a series of maps of the planet, from the deepest trenches in the oceans to the ice at the poles. Users see and hear scientists explain how ice, the oceans, precipitation, and temperatures have changed over time.

The **Minerals Education Coalition** (MEC) of the **Society for Mining, Metallurgy, and Exploration,** an AGI member society, supported Earth Science Week in 2017. MEC provided a poster featuring the educational "What's in My Toothpaste?" mining-and-minerals game for the 2017 educator kit. Earth Science Week's 2017 activity calendar featured a "Paste With a Taste" classroom activity from MEC. And program participants received information about a series of MEC podcasts with industry experts, showing students how an interest in STEM subjects can lead to a rewarding career in the mining industry.

SUMMARY OF ACTIVITIES
• KEY PARTNERSHIPS AND EFFORTS

Earth Science Week directed participants' attention to a joint effort by AGI and the **National Association of Geoscience Teachers** (NAGT) to strengthen implementation of the Next Generation Science Standards (NGSS) at the state level. Following an NGSS Summit of education leaders in April 2015, science educators have regularly participated in free webinars and joined discussions online. In addition, Earth Science Week participants have been invited to take advantage of NAGT offerings including online lessons, NAGT's Outstanding Earth Science Teacher Awards, the Dorothy Stout Professional Development Grants, and the Journal of Geoscience Education.

NASA, a founding partner of Earth Science Week, provided a folder of materials in the Earth Science Week 2017 Toolkit (and online) designed to frame phenomena-based student investigations. Three of the five pages in the packet list examples of NASA resources divided into grade bands: elementary, middle, and high school. Each resource not only aligns to the event theme of "Earth and Human Activity," but also supports two essential ideas of NGSS: ESS2: Earth Systems and ESS3: Earth and Human Activity. The other two pages lead educators to a better understanding of accessing and using NASA data and images. Earth Science Week's 2017 activity calendar also featured a "Carbon Travels" classroom activity from NASA. Throughout the year, Earth Science Week promoted awareness of NASA's online offerings, such as NASA Wavelength, SciJinks, Space Place, and Climate Kids.

The **National Cave and Karst Research Institute** provided an informational bookmark for the Earth Science Week 2017 Toolkit.

The **National Earth Science Teachers Association** (NESTA), a longtime Earth Science Week partner, continued its vital role in helping AGI promote excellence in geoscience education. At the National Science Teachers Association Annual Conference, the NESTA Reception included a ceremony during which a teacher was given the Edward C. Roy, Jr. Award for Excellence in K-8 Earth Science Teaching. NESTA members also received copies of the Earth Science Week 2017 poster in their association newsletter. Finally, NESTA assembled an incredible Earth Science Week Collection of teaching resources for educators. Day by day, the collection offered instructional resources tailored for each of Earth Science Week's Focus Days, from Sunday's International EarthCache Day through Saturday's International Archaeology Day. In addition, participants were invited to join the NESTA community at the close of Earth Science Week for a virtual celebration on Saturday, October 14.

Earth Science Week raised awareness of **National Environmental Education Week** (EE Week), the nation's largest environmental education event. Focusing largely on STEM topics, EE Week connected educators with resources to promote K-12 students' understanding of the environment.

Earth Science Week promoted the **National Groundwater Association**'s (NGWA) Groundwater Awareness Week in March 2017 as well as NGWA's Protect Your Groundwater Day program in September 2017, advocating water conservation and safety. The AGI member society offers Groundwater Adventures, a website providing educational activities for young people.

Earth Science Week's 2017 activity calendar featured a "Citizen Science" activity from the **National Oceanic and Atmospheric Administration** (NOAA). In addition, Earth Science Week 2017 directed participants to NOAA's online multimedia education resources, including lesson plans, real-world data, instructional games, videos, and more.

A longtime Earth Science Week partner, the **National Park Service** (NPS) continued for the eighth year a major component to its involvement in Earth Science Week. National Fossil Day, established as a celebration to take place annually on the Wednesday of Earth Science Week, once again reached millions of people. Earth Science Week promoted awareness of NPS's The interactive Web Ranger program, which helps people of all ages learn about the national parks. Program participants also learned about the NPS National Natural Landmarks program, which recognizes and encourages the conservation of sites that contain outstanding biological and geological resources. NPS videos on climate change were made available to program participants. Posters illuminating the geologic resources relating to glaciers, atmospheric resources relating to views of the parks, and fossil resources of National Natural Landmarks, appearing in the Earth Science Week 2017 Toolkit, successfully continued the series of park posters produced collaboratively by the federal agency and AGI. The agency also served as a partner in the Geologic

SUMMARY OF ACTIVITIES
• KEY PARTNERSHIPS AND EFFORTS

Map Day initiative in 2017. Earth Science Week's 2017 activity calendar also featured a classroom activity on "Parks Past, Present, and Future."

Earth Science Week participants learned about soil education resources offered online by the U.S. Department of Agriculture's **Natural Resource Conservation Service** (NRCS). Resources for the elementary level include lesson plans, links to soil education websites, and even soil songs. For this year's Earth Science Week Toolkit, the Natural Resources Conservation Service provided a 2017 Soils Planner, featuring month-by-month information on soil science.

Earth Science Week partnered with the **National Science Teachers Association** (NSTA) again in 2017, reaching science educators nationwide. Program participants learned about "Freebies for Science Teachers" on the NSTA website. Also, AGI exhibited once again at the NSTA Annual National Conference, sharing Earth Science Week and other education material for science teachers.

The **National Wildlife Federation** provided an informational postcard for the Earth Science Week 2017 Toolkit, providing details and links for an Eco-Schools USA and GLOBE collaboration to foster STEM education.

The Nature Conservancy, which offers informational resources ideal for educators aiming to teach about a wide range of geoscience topics, was promoted through Earth Science Week. Videos, webcasts, articles, and other materials conveyed the work of scientists engaged in conservation efforts around the world.

Gearing up for National Fossil Day, Earth Science Week directed program partners' attention to the **Paleontological Research Institution** (PRI), an AGI member society providing education materials and opportunities for science teachers and students at all grade levels. The online Teacher Friendly Guide, for example, gives brief geologic histories of every region of the United States.

Partners in Resource Education (PRE), a longtime program partner, provides activities focusing on the geoscience of conservation. The consortium of seven federal agencies educates thousands of young people, introduces them to natural resource careers, and cultivates the next generation of land and water stewards. In 2017, PRE collaborated to promote awareness of Earth Science Week, and vice versa.

The **Public Broadcasting Service** (PBS) partnered with Independent Television Service Inc. and WGBH to deliver a three-part virtual professional development series focused around climate change and

Earth Science Week 2017 Photo Contest entry by finalist Taylor Russon

climate science, as Earth Science Week participants learned. The series offered support to teachers across the country who are struggling to overcome preconceived notions around climate change and climate science, which are a part of the middle and high school Next Generation Science Standards. This series of virtual professional development, consisting of one webinar and two television events, introduced educators to high-quality, science- and media-based educational materials.

Schlumberger Excellence in Educational Development (SEED) is a nonprofit education program that empowers educators to share their passion for learning and science with students. In addition to promoting awareness of SEED and other resources, AGI has partnered with the program to provide geoscience education resources in both Spanish and English since 2010.

National Public Radio's **Science Friday** reached some 4.5 million people with its radio broadcast, online content, and social media promoting Earth Science Week. During the popular call-in talk show, host Ira Flatow launched a Science Friday Science Club inviting listeners to share favorite rocks online in a "Neat Rock Challenge." People nationwide shared their #**neatrock** and learned about Earth science. Participants were linked to informative articles, scientific expertise, and a webinar on rock collections.

Earth Science Week partnered with Denver's **Scientific and Cultural Facilities District** (SCFD) to provide

SUMMARY OF ACTIVITIES
• KEY PARTNERSHIPS AND EFFORTS

materials to teachers at Educator Night at the Denver Museum of Nature and Science, where more than 1,900 teachers and guests enjoyed exhibits and learned educational improvement strategies.

For teachers aiming to "shake up" education, Earth Science Week shone a spotlight on the **Seismological Society of America** (SSA). SSA's website provided seismic eruption models, wave animations, plate tectonics simulations, information on tsunamis, and more. SSA also offered publications, information on seismology careers, a distinguished lecturer series, and an electronic encyclopedia of earthquakes.

Program participants received information about **Smithsonian Education,** which offers a fascinating exploration of Earth's soil with its "Dig It! The Secrets of Soil" exhibition. Information, videos, expert instruction, and activity sheets are available online.

Earth Science Week participants were encouraged to take advantage of offerings of the **Society of Exploration Geophysicists** (SEG), which provides programs for educators and students. For example, the distinguished lecturer series and honorary lecturer series both enabled students to meet professional geophysicists, learn about groundbreaking research in the field of seismic research, and obtain valuable career information.

Advanced by the **Society of Petroleum Engineers** (SPE), the Energy4Me program offers teachers a collection of tools for teaching about oil, gas, and other energy sources, including classroom activities, experiments, and presentations, as well as teacher workshops and energy education materials for the classroom.

The **Soil Science Society of America** (SSSA), a longtime program partner, provides lessons, activities, fun facts, sites of interest, and soil definitions for the novice soil scientist online. These resources were promoted by the October event. Earth Science Week's 2017 activity calendar featured a "Liquefaction" classroom activity courtesy of SSSA. In addition, SSSA provided a magnetized "Soil!" bookmark that was included in the educator kit.

Program participants learned about a new Earth Science Week partner, **STEMIE,** an education framework that elevates youth invention and entrepreneurship education to a core part of K-12 education. STEMIE stands for Science, Technology, Engineering, and Math linked to Invention and Entrepreneurship (STEM+I+E) and maps essential unstructured problem-solving teaching activities to core STEM curricula and standards. On the website, educators learned more about the organization's framework, coalition, events, resources,

and how students can get involved. AGI staff participated in judging geoscience entries in STEMIE's annual student competition.

SWITCH Energy Alliance provided an informational postcard on this video-based energy education project, including online resources, for the Earth Science Week 2017 Toolkit.

The Earth Science Week program ended the year by announcing the next year's program theme, "Earth as Inspiration," with great fanfare at the **Torpedo Factory Art Center** in Alexandria, Virginia. Teachers, students, and others explored Earth science alongside the arts during "The Late Shift: STEAM-Powered December." The event featured hands-on activities, demonstrations, food, music, and more. AGI Earth Science Education Ambassador and former U.S. Secretary of the Interior Sally Jewell spoke on the importance of geoscience in society.

For the Earth Science Week 2017 Toolkit, **UNAVCO** provided a poster on "Tectonic Motions of Alaska," including detailed information on the GPS technology and data used to study these phenomena

The **U.S. Bureau of Land Management** (BLM), a continuing Earth Science Week partner, provided a flyer including a dinosaur activity sheet for the 2017 educator kit. BLM also was the subject of Earth Science Week promotions, including its Classroom Investigation Series online. The Earth Science Week 2017 activity calendar included a "Groundwater on the Move" activity from BLM.

The **U.S. Department of Energy**'s Office of Energy Efficiency & Renewable Energy website offered classroom activities and materials for K-12 science instruction, as program participants learned. Additionally, educators were invited to explore DOE's websites for the National Renewable Energy Laboratory. Earth Science Week participants were made aware that AGI's Center for Geoscience & Society produced education materials, including videos in English and Spanish, education guides, a "quick start" guide to energy literacy, lesson connections, and guidance on aligning energy literacy lessons with the Next Generation Science Standards. Essential Principles and Fundamental Concepts for Energy Education resources were made available on the DOE website. Teams of program participants were urged to enter DOE's new Geothermal Design Challenge, which invites young people to design an infographic that illustrates how geothermal energy is clean, safe, reliable, and renewable. DOE also provided a STEM Spark bookmark for the Earth Science Week 2017 Toolkit.

SUMMARY OF ACTIVITIES
• KEY PARTNERSHIPS AND EFFORTS

Earth Science Week also promoted awareness of the **U.S. Environmental Protection Agency**'s collection of free resources to enhance middle school students' understanding of climate change impacts on the United States' wildlife and ecosystems. The online toolkit includes case studies, activities, and videos based on climate science, environmental education, and stewardship information.

Overlapping Earth Science Week 2017, National Wildlife Refuge Week was held October 9-15. The event, celebrating the richness of the 550 units that make up America's National Wildlife Refuge System, was sponsored once again by the **U.S. Fish and Wildlife Service** (FWS), an Earth Science Week partner.

Earth Science Week participants learned about online education resources offered by the **U.S. Geological Survey** (USGS), a longtime Earth Science Week partner and supporter, as well as the thousands of free images and over 69,000 searchable publications such as maps, books, and charts provided online by the agency. Earth Science Week's 2017 activity calendar featured a "Measuring Glacial Retreat" classroom activity courtesy of USGS. Also, USGS continued its leadership role as a founding partner of Geologic Map Day in 2017, providing support as well as its National Geologic Map Database's MapView, which offers a mosaic view of published geologic maps. During Earth Science Week, USGS and AGI collaborated to hold a "Geologic Open House" at Great Falls Park in Virginia. Attendees learned about the area through talks and guided tours led by USGS geoscientists.

Earth Science Week partnered with **Washington Union Station** to exhibit over 20 top photos entered in the program's 2017 photo contest, which explored the theme of "Earth and Human Activity Here." Located in the heart of the nation's capital, Union Station is one of the country's busiest train stations, with an average of about 100,000 train and local metro passengers passing through daily.

Education resources of the **Water Environment Federation** (WEF) were promoted among Earth Science Week participants, especially during World Water Day in March, program participants learned about clean water and real-life professionals who keep water resources safe. WEF is a nonprofit association that provides technical education and training for water quality professionals.

Earth Science Week participants in Canada were invited to take part in the 2017 **WHERE Challenge** sponsored by Teck Resources Limited. WHERE stands for the places where Earth scientists work: Water, Hazards,

Earth Science Week 2017 Visual Arts Contest entry by finalist Daniel Chia

Energy, Resources, and Environment. The challenge, a national contest endorsed by the Canadian Earth sciences community, asked students ages 9-14: "What on Earth is in your stuff, and WHERE on Earth does it come from?" Students explored the nonrenewable Earth resources that make up their favorite objects.

Earth Science Week promoted awareness of **Windows on Earth,** an online educational project that features photographs taken by astronauts on the International Space Station. The site is operated by TERC, an educational non-profit, in collaboration with the Association of Space Explorers (the professional association of flown astronauts and cosmonauts), the Virtual High School, and CASIS (Center for Advancement of Science in Space). The images help show Earth from a global perspective.

PLAN!T NOW's **Young Meteorologist Program** took students on a severe weather preparedness adventure while learning about severe weather science and safety. Promoted by Earth Science Week, the program was developed in partnership with NOAA's National Weather Service and the National Education Association. Young Meteorologists were given opportunities to put their knowledge to work in hands-on activities and community service projects.

SUMMARY OF ACTIVITIES
• EARTH SCIENCE WEEK TOOLKITS

Earth Science Week Toolkits

Across the country, AGI distributed some **14,000 Earth Science Week Toolkits** to teachers and geoscientists in 2017. The number of AGI member societies requesting complimentary Earth Science Week Toolkits for distribution was 18, and the number of state geological surveys requesting complimentary kits for distribution was 29.

As in past years, thousands of kits also were distributed through program partners including USGS, NASA, the National Park Service, and AAPG Student Chapters. Hundreds of kits were shipped free to geoscience department chairs at colleges and universities nationwide. Toolkits were shipped to program participants and around the world.

The 2017 toolkit featured AGI's traditional Earth Science Week poster, education and outreach flyer, and school-year calendar showcasing geoscience classroom investigations and important dates of Earth science events.

For the first time, the school-year calendar's classroom investigations featured notations explaining to educators exactly how each activity aligns with expectations outlined in the **Next Generation Science Standards.** Additionally, program partners' contributions, many of which also included activities aligned with the standards, made the Earth Science Week 2017 Toolkit one of the richest in recent years.

Web Resources

According to Google Analytics, the **Earth Science Week website** was accessed by users in **213 nations, territories, and regions** worldwide in 2017. The program website (**www.earthsciweek.org**) delivers essential resources for educators throughout the year. As in past years, the Earth Science Week website was updated regularly to reflect the new theme, contests, proclamations, events, initiatives, and classroom activities for 2017. The entire site received more than 633,000 page views. Within the site, Classroom Activities pages received over 427,000 views. Contests pages received more than 31,000 views. Plan an Event pages received over 6,100 views.

New to the Earth Science Week 2017 Photo Contest, AGI and Earth Science Week debuted an online resource, the **"Earth and Human Activity Here" Photo Map.** Select entries are featured on the map, linked to the location of origin. This innovation serves as a powerful educational resource, fueling discussions in classrooms and other settings. As contest entries arrived, staff populated the global map with examples of ways that people interact with Earth systems where they live. Program participants, students, and teachers were invited to view the wide variety of forms those interactions take.

Those hosting events during Earth Science Week 2017 were invited to let people know about it at **Events in Your Area.** This web page provided information on events taking place through program partners in each state, such as exhibits, tours, lectures, and open houses. The new **Earth Science Week Event Registry** enabled participants to promote their events more effectively than ever. All registered events were listed on Earth Science Week's Events In Your Area site.

In addition, participating groups could be listed in **Earth Science Organizations,** an online map that offers clickable links to Earth Science Week events taking place at parks, museums, science and technology centers, university geology departments, local geological societies, and other nearby locations.

Promoted through various online channels, AGI's **Earth Science Week promotional video** trumpeted the importance of the geosciences and the celebration's role in promoting public awareness. This brief, exciting, eye-popping video answers key questions: Why is Earth science a big deal? How does Earth Science Week help promote learning and teaching about the subject? And what can students, educators, community partners, and others do to get involved?

While exploring the Earth Science Week theme of "Earth and Human Activity," science teachers and students were invited to consider the ways that people worldwide recently have begun to recognize, designate, and conserve special places and natural resources that represent our rich geologic heritage, or "geoheritage." A new educational resource of the Earth Science Week website, the **Our Shared Geoheritage** page, features educational material on this heritage and links users to recommended resources, including downloadable reports, articles, blogs, geoheritage locations, and learning activities. The page also features geoheritage-related classroom activities and links to information on geoheritage in every state.

SUMMARY OF ACTIVITIES
• NEWSLETTER

Another recent addition is an Earth Science Week page helping students explore human interaction with Earth systems through visualizations. The **Visualizing Earth Systems** page links users to dozens of recommended visualizations dealing with energy, climate, minerals, water, hazards, and other topics. In addition, the page links users to overviews of these topics provided by AGI's Critical Issues Program.

Once again AGI offered four quarterly **Earth Science Week Webcasts** in 2017, expanding the program's use of online formats and media for public outreach. The free webcasts provided lively overviews of Focus Days (spring), Contests (summer), the Toolkit (autumn), and the Roy Award (winter). Each roughly five-minute tutorial includes a wealth of online links, which viewers can click during the narrated presentation to review available resources.

Program participants were encouraged to visit the continually updated Earth Science Week **Classroom Activities** page for more than 130 free learning activities, most of them contributed by leading geoscience agencies and groups. Activities are organized and searchable by various criteria, including specific Earth science topics. To find the perfect activity for a lesson, teachers can search by grade level and science education standard. Maybe most useful, they also can search among 24 categories of Earth science topics, from energy and environment to plate tectonics and weathering. This may be why Classroom Activities rank as one of the program's most popular online offerings, with 94 percent of survey respondents rating it as "useful" or "very useful."

AGI provided a set of free online videos and other electronic resources to help students, educators, and others explore the "big ideas" of Earth science during Earth Science Week 2017 and throughout the year. **Big Ideas Videos** bring to life the nine core geoscience concepts that everyone should know. The Earth Science Literacy Initiative, funded by the National Science Foundation, codified these principles. The videos are available on

Earth Science Week 2017 Photo Contest entry by Tanvi Dhaka

YouTube and TeacherTube. The Earth Science Week website also provides dozens of classroom activities linked to the "big ideas."

A page dedicated to **Geoscience Career, Scholarship, and Internship Resources** was added to the program website. Another page of links was included to provide external connections to sites featuring resources on key topics such as chemistry, climate, drought, earthquakes, energy, floods, hurricanes, landslides, sinkholes, soil, tornadoes, tsunamis, volcanoes, and wilderness fires.

Finally, Earth Science Week makes ample use of online social networking to reach new audiences, especially young people. The program's presence on **Facebook,** the Internet's most popular networking site, includes an Earth Science Week Fan Page. In addition, web surfers are invited to receive geoscience news, resources, and opportunities by following Earth Science Week on **Twitter.** Tweets are sent frequently, whenever there is valuable news or information to share. The number of people learning about Earth Science Week through social media remained impressive in 2017, as more than 155,000 people received program information from AGI and program partners through Facebook, Twitter, and Pinterest.

Newsletter

The monthly **Earth Science Week Update** newsletter reached some 6,000 teacher, student, and geoscientist subscribers in the past year. The electronic newsletter kept planners and participants up-to-date on Earth Science Week planning at the national level, encouraged participation in local areas, and provided news on geoscience topics of interest to participants.

Besides highlighting worthwhile resources, these monthly e-mail updates reinforce the belief that geoscience education is a priority throughout the year, not only during one week each October. It is little wonder that the e-newsletter remains a popular online offering, with virtually 100 percent of survey respondents rating it as "useful" or "very useful."

SUMMARY OF ACTIVITIES
• CONTESTS

Earth Science Week 2017 Photo Contest winning entry by Roxie Khalili

Contests

AGI held contests in connection with Earth Science Week for the 17th consecutive year. Contests were designed to encourage K-12 students, teachers, and the general public to become involved in the celebration by exploring artistic and academic applications of Earth science. Earth Science Week expanded eligibility for its photo contest to allow international members of **AGI Member Societies** and **AGI International Affiliates** to participate.

Four contests continued to provide ways for many people to participate in Earth Science Week. Photos, art, videos, and essays were submitted by hundreds of people. Each first-place winner received $300 and a copy of AGI's *The Geoscience Handbook*. Entries submitted by winners and finalists were posted online.

David De Costa of Alexandria, Virginia, won first place in the visual arts contest with a creative and colorful drawing of earth, water, air, and living things. Finalists were Ian Lee, Vinuth Sumanasiri, Daniel Chia, and Pranavi Chatrathi. Students in grades K-5 made two-dimensional artworks illustrating the theme **"People and the Planet."** Roxie Khalili of Foster City, California, won first place in the photo contest with an image of human-built infrastructure surrounding the lagoon and its inhabitants. Finalists were Taylor Russon, Andrew DePriest, Jill Holz, and Harrison Cho. Submissions illustrated the theme **"Earth and Human Activity Here."** Tracy Peucker of Virginia Beach, Virginia, won first place in the essay contest with a paper on "The Effect of Geosciences on Landslide Prevention." Finalists were Ellie Kain-Kuzinewski, Hannah Shin, Madeline Marous, and Ryder Robins. Students in grades 6-9 wrote essays of up to 300 words addressing this year's theme, **"Human Interaction With Earth Systems."** Sophi Schneider

Earth Science Week 2017 Visual Arts Contest entry by the winner, David De Costa

Below, part of the winning entry to AGI's 2017 Earth Science Week essay contest by Tracy Peucker
See all the winners and read the rest of Tracy's essay at **https://www.earthsciweek.org/contests/2017**

The Effect of Geosciences on Landslide Prevention by Tracy Peucker

Landslides cause three billion dollars worth of damage, and 1,000 deaths each year throughout the world. Entire neighborhoods, towns, even cities are destroyed, though dams are most commonly destroyed (USGS, 2005). Geoscientists study landslide prone areas to find preventative solutions based upon the causes, both artificial and naturally occurring. By implementing geoscience research, landslides could be greatly decreased, as well as predicted.

Landslides often occur during the monsoon season, or following earthquakes. Heavy rains can cause loose soil, beginning the movement of debris. Human influence concerns include unsafe piping, or excess irrigation (USGS, 2016). Mining and building can lead to landslides, specifically near unstable land. Earthquakes often produce unstable land, tremors can trigger landslides with almost no warning. Unstable land can also stem from deforestation, caused by wildfire, disease, or industry. Both natural causes and human infrastructure have equal influence upon landslides occurrence.

SUMMARY OF ACTIVITIES
• EARTH SCIENCE TEACHER AWARD

won first place in the video contest with her video on "Our Beautiful Earth." Finalists were Tahtinen's Firsties and Kolby Noble. Individuals and teams created brief, original videos that tell viewers how people affect Earth systems, or how Earth systems affect people, through **"Earth Connections"** in their part of the world.

In cooperation with Washington, D.C.'s Union Station, Earth Science Week staged a **"Earth and Human Activity Here" Photo Exhibit** featuring select images from in the 2017 photo contest. The historic facility is visited by some 100,000 train and local metro passengers passing through daily. The partnership launched what is expected to be an ongoing program offering, exhibits of photos in high-traffic public spaces nationwide.

AGI's Geoff Camphire and M.J. Tykoski, 2017 winner of the Edward C. Roy, Jr. Award at the NESTA reception during NSTA's 2017 conference
Image credit: Courtesy of Geoff Camphire

Earth Science Teacher Award

For the 10th consecutive year, AGI and the AGI Foundation offered the **Edward C. Roy, Jr. Award for Excellence in K-8 Earth Science Teaching.** The 2017 award went to **M.J. Tykoski,** an eighth-grade teacher at Cooper Junior High School in Wylie, Texas. Tykoski earned her master's degree in educational leadership from Grand Canyon University in Phoenix, Arizona. She is a member of several professional organizations, including the Texas Earth Science Teachers Association, and is the recipient of numerous grants and other awards in education.

Earth Science Week 2017 Visual Arts Contest entry by finalist Pranavi Chatrathi

For the first time, AGI produced a brief **Roy Award video** honoring the winner and promoting the importance of high-quality education in the geosciences. Tykoski also received a $2,500 prize and an additional grant of $1,000 to enable her to attend the National Science Teachers Association 2017 National Conference to accept the award during a reception hosted by the National Earth Science Teachers Association. Finalists for the award were Troy J. Simpson of Glenn Raymond School in Watseka, Illinois, and Chad Pavlekovich of Salisbury Middle School in Salisbury, Maryland.

The award recognizes one classroom teacher from kindergarten to eighth grade for leadership and innovation in Earth science education. This award is named in honor of Dr. Edward C. Roy, Jr., a past president of AGI and strong supporter of Earth science education. In addition to U.S. teachers, instructors throughout the United Kingdom were invited to compete for the prize. U.K. teachers were provided with detailed guidance on entering the competition by AGI and **The Geological Society of London**, a member society and Earth Science Week partner.

Earth Science Week 2017 Photo Contest finalist entry by Jill Holz

SUMMARY OF ACTIVITIES
• FOCUS DAYS

Focus Days

Earth Science Week 2017 kicked off on Sunday, October 8, with the ninth annual **International EarthCache Day**. "EarthCaching" is a variation of a recreational activity known as geocaching, in which a geocache organizer posts latitude and longitude coordinates on the Internet to advertise a cache that geocachers locate using GPS devices. The activity has attracted over a million participants worldwide. When people visit an EarthCache, they learn something special about Earth science, the geology of the location, or how the Earth's resources and environment are managed there. EarthCaching has been developed by the Geological Society of America, a major program partner.

On Monday, October 9, educators and young people were encouraged to explore "big ideas" as part of **Earth Science Literacy Day**. The AGI "Big Ideas of Earth Science" videos provided on YouTube and TeacherTube outline the core concepts of geoscience, as codified by the Earth Science Literacy Initiative with support from the National Science Foundation. To help teachers and students use the videos, the Earth Science Week website offers dozens of related classroom activities.

One of the highlights of recent years' Earth Science Week celebrations has been **"No Child Left Inside" Day**, an event that in its inaugural year engaged some 500 students in outdoor learning activities and received coverage by news media from NBC to NPR. In 2017, students and educators nationwide were invited to take part on the Tuesday of Earth Science Week, October 10. AGI's online NCLI Day Guide provided everything needed to plan a local NCLI Day event. The free guide provided 17 outdoor activities, as well as detailed recommendations for creating partnerships, planning logistics, reaching out to the local media, and following up in the classroom.

In addition, on the Tuesday of Earth Science Week 2017 participants were invited to take part in **Earth Observation Day**. Previously celebrated at other times of the year, this October 10 event aimed to engage students and teachers in remote sensing as an exciting and powerful educational tool. The event was a STEM educational outreach event of AmericaView and its partners. AmericaView is a nationwide partnership of remote sensing scientists who support the use of Landsat and other public domain remotely sensed satellite data through applied remote sensing research, K-12 and higher STEM education, workforce development, and technology transfer. Participants made use of lessons and activities by AmericaView and other organizations, as well as additional Earth Observation Day resources, online.

AGI Education and Outreach staff at the 2017 National Fossil Day event
Image credit: AGI/J. Lilek

Earth Science Week featured the return of a popular event on Wednesday, October 11, 2017. In partnership with the National Park Service (NPS), AGI helped conduct the eighth annual **National Fossil Day,** including activities and resources designed to celebrate the scientific and educational value of fossils, paleontology, and the importance of preserving fossils for future generations. NPS offered a website full of educational resources and information designed specifically for students and teachers. On the site's NPS Fossil Park Highlights page, visitors could find lesson plans developed to reflect state standards, fossil trading cards, videos about pygmy mammoths, special brochures, a virtual museum exhibit on dinosaurs, and more. NPS also held a National Fossil Day Art Contest. Finally, AGI collaborated with NPS partners and other geoscience organizations to conduct a National Fossil Day event in Washington, D.C. Fossil enthusiasts in Washington, DC celebrated National Fossil Day on the National Mall from, where AGI educated and entertained visitors with a greenscreen-equipped "Paleontology Play Space" photo booth on the pathway between the Smithsonian Castle and the National Museum of Natural History. Visitors, who had their pictures taken virtually in the midst of amazing fossil finds, received photo souvenirs of the day.

Program participants were invited to join the Earth Science Week team in encouraging everyone — including women, minorities, and people with a range of abilities — to explore geoscience careers on **Geoscience for Everyone Day,** Thursday, October 12. Educators welcomed geoscientists into the classroom to speak. Geoscientists visited schools and volunteered at science centers. Others organized scout events, led field trips, and held special "Take Your Child to Work Day" events. The aim was to open a young person's eyes to the world of Earth science. Doing so, participants supported the efforts of AGI member societies such as the Association for Women Geoscientists and the National Association of Black Geoscientists in raising awareness of the remarkable opportunities available to all young people in the

SUMMARY OF ACTIVITIES
• SPECIAL EVENTS

Earth sciences. The program website directed participants to "Visiting Geoscientists: An Outreach Guide for Geoscience Professionals," a handbook co-produced by AGI and the American Association of Petroleum Geologists' Youth Education Activities Committee.

The sixth annual **Geologic Map Day** held on Friday, October 13, 2017, promoted awareness of the study, uses, and importance of geologic mapping for education, science, business, and a variety of public policy concerns. The final event for the school week of Earth Science Week 2017 was hosted by the U.S. Geological Survey and the Association of American State Geologists in partnership with AGI, along with additional partners including the National Park Service, the Geological Society of America, and NASA. Students, teachers, and the wider public tapped into the various educational activities, print materials, online resources, and public outreach opportunities for active participation. The Earth Science Week 2017 Toolkit contained a Geologic Map Day poster that provided geologic maps, plus step-by-step instructions for a related classroom activity on "Karst, Sinkholes, and Human Activity." Additional resources for learning about geologic maps were featured on the Geologic Map Day web page of the Earth Science Week site. Activities nationwide, many led by state geologic surveys, spurred learning in schools. On the preceding day, Earth Science Week, in collaboration with the U.S. Geological Survey, celebrated Geologic Map Day with a "Geologic Open House" at Great Falls Park in Virginia. Attendees not only learned about the natural forces that

AGI's Juliet Crowell with costumed museum educators at Educators' Night in Denver
Image credit: AGI/Juliet Crowell

shaped this landscape over billions of years, but explored the park's rocky terrain, waterfalls, and more in guided tours led by USGS geoscientists.

Earth Science Week 2017 reached its climax with **International Archaeology Day** on Saturday, October 14. The event was a celebration of archaeology and the thrill of discovery. Every October, archaeological programs and activities for people of all ages and interests are presented by the Archaeological Institute of America and archaeological organizations across the United States, Canada, and elsewhere. Programs included activities such as a family-friendly archaeology fair, a guided tour of a local archaeological site, a simulated dig, and a lecture or a classroom visit from an archaeologist. In every case, interactive, hands-on International Archaeology Day programs provided the chance for participants to indulge their inner "Indiana Jones."

Special Events

Earth Science Week 2017 made a big noise about geoscience with the help of National Public Radio's popular call-in talk show, **Science Friday.** Award-winning host Ira Flatow kicked off a Science Friday Science Club that invited listeners and geoscience enthusiasts to share their favorite rocks online in a "virtual rock collection" Neat Rock Challenge. Science Friday's Earth Science Week effort reached an estimated 4.5 million people by radio, website, and social media.

Throughout October, AGI geoscientists and other experts chatted with people from across the country who shared their **#neatrock**, delved into the rock's story, and learned about the amazing world of geoscience. The Science Club connected participants with informative articles, scientific expertise, and a webinar on ways that students and educators could explore their rock collections together. Webinar participants discussed strategies for rock-inspired inquiry with a team of experienced

geoscience experts and educators, who explored ways to engage with a child's enthusiasm, existing knowledge, and ability to observe, describe, and make inferences about rocks.

City-specific celebrations served as major centers of public awareness activities during Earth Science Week 2017. **Houston, Denver, Washington, D.C., and Richmond, Virginia** extended and deepened the reach of the nationwide event. In all these cities, AGI donated hundreds of Earth Science Week Toolkits and collaborated with geoscience organizations and public schools to provide special events, educational materials, online resources, and activities in schools and other settings.

AGI's **Citywide Celebrations website** provided educators, students, and families with links to additional educational resources as well as other offerings in participating cities. Program participants nationwide were

SUMMARY OF ACTIVITIES
• AGI PROMOTIONS

encouraged to collaborate with local partners to launch their own Citywide Celebration.

AGI staff exhibited at the **Energy Day Houston** festival as part of its new collaboration with Consumer Energy Alliance. Earth Science Week shared geoscience-oriented STEM materials with the many of people who visited its booth at this free outdoor event, which for years has showcased education technologies and innovations. The event also underscored the impact and geoscientific relevance that Hurricane Harvey had on Houston-area schools, families, and communities. Organizers estimated roughly 30,000 people attended the event.

Additionally, Earth Science Week exhibited at the first-ever **Energy Day Colorado**, a free family festival showcasing interactive demonstrations and exhibits across science, technology, engineering, and mathematics (STEM). This Denver event invited students and their families to engage with industry experts and educators from a variety of organizations, helping to spark an interest in energy, environmental, and STEM careers for K-12 students. Participants enjoyed music, food, and fun educational activities.

While in Denver, AGI staff also exhibited at **Educator Night** at the Denver Museum of Nature and Science, where more than 1,900 teachers and guests received Earth Science Week materials and viewed exhibits.

Earth Science Week finished 2017 with a major event to kick off the next year's program theme, "Earth as Inspiration." AGI welcomed hundreds of educators, students,

AGI Earth Science Education Ambassador Sally Jewell at AGI's "Earth as Inspiration" event in Alexandria, Virginia
©Laura Hatcher Photography

and the general public to explore connections between the geosciences and the arts during **"The Late Shift: STEAM-Powered December"** at the Torpedo Factory Art Center in AGI's headquarters city of Alexandria, Virginia.

The free event included hands-on activities, demonstrations linking art and geoscience, musical performances, and opportunities to craft "take home" artworks based in Earth science. Additionally, educational materials such as Earth Science Week Toolkits were distributed to teachers, students, homeschoolers, and others. Roy Award winner M.J. Tykoski was formally recognized for excellence in Earth science teaching. A highlight of the evening was a series of addresses on the importance of geoscience by **AGI Earth Science Education Ambassador and former U.S. Secretary of the Interior Sally Jewell, Alexandria Mayor Allison Silberberg, and AGI Executive Director Allyson Anderson Book.**

AGI Promotions

Earth Science Week promoted awareness of numerous AGI programs and resources of interest to Earth science educators, students, and enthusiasts, including AGI's Center for Geoscience & Society, the AGI Geoscience Workforce program, the William L. Fisher Congressional Geoscience Fellowship, AGI's NGSS Education Webinars, the Pulse of Earth Science website, the Visiting Geoscientists guide, AGI's Critical Issues Program, Geoscience in Your State Factsheets, the Earth Science Organizations website, the Faces of Earth DVD, EARTH magazine, AGI Information Services (such as GeoRef), District Visit Days, the Why Earth Science video, and AGI's and the National Park Service's jointly published *America's Geologic Heritage: An Invitation to Leadership*.

Congressional Recognition

In a display of bipartisanship, members of the U.S. House of Representatives from across the country came together to introduce a resolution expressing support for designation of the week of October 8-14, 2017, as Earth Science Week.

Rep. Jared Polis (D-Colorado) submitted H. Res. 556, on behalf of himself and Reps. Barbara Comstock (R-Virginia) and Dan Lipinski (D-Illinois) on Oct. 10, 2017. Recognizing Earth Science Week 2017 as the 20th annual celebration of the signature event organized by the AGI, the resolution noted that "the study of Earth sciences leads to an improved understanding of the Earth's natural systems and the interplay between human society and these systems."

SUMMARY OF ACTIVITIES

• STATE PROCLAMATIONS

The "Green Screen" activity at AGI's "Earth as Inspiration" event in Alexandria, Virginia
©Laura Hatcher Photography

> ### State Proclamations
>
> Seven states have demonstrated outstanding science-literacy leadership by issuing **"perpetual proclamations"** of Earth Science Week, ensuring recognition every year: Alaska, Delaware, Illinois, Nevada, North Dakota, Oklahoma, and South Dakota.
>
> Governors also issued single-year proclamations in seven additional states — Arkansas, Maryland, Missouri, Nevada, Oregon, Pennsylvania, and Tennessee — bringing the total number of states with proclamations of recognizing Earth Science Week 2017 to 14.

Publicity and Media Coverage

AGI enlisted the support of a wide range of media in promoting awareness of Earth Science Week, resulting in unprecedented reach for promotional activities in 2017 and helping to lay a foundation for more coverage in years to come.

Earth Science Week 2017 news, events, programs, and resources were covered by **national news organizations** such as *AAPG Explorer,* American Geophysical Union, *Altenergymag.com, American News,* Archaeological Institute of America, Bahrain News Agency, Before It's News, *Congress.gov,* Consumer Energy Alliance, *The Denver Post, Earth, The Earth Scientist, Environment Guru, Eos, GSA Today,* Hands on the Land, *Highly Allochthonous, Hot Travel News, The Leading Edge, Life Magazine* Blog Spot, *The Mining Journal,* NASA, National Park Service, *National Science Foundation News,* Newstral, *NXTE News,* Omni Innotech, Opportunity Desk, PLOS Paleo Community Blog, Public Now, Reddit, *RobinsPost.com, Science Friday* of National Public Radio, *Scoopnest,* University of Cambridge in the United Kingdom, *USA Online Journal,* U.S. Geological Survey, and *Wopular.*

Additionally, the event was covered by **international news organizations** including *Advanced Science News* of Wiley-VCH in Germany, *BCTV News* of Canada, *Breaking Belize News* in Belize, *The Daily Mail* of the United Kingdom, *DurhamRegion.com* of Canada, *Earth Touch News Network* of South Africa, *Financial Tribune* of Iran, *The Guardian* of England, *Mogaz News* in England, *Newsround* of Australia, *The Northern Times* of the United Kingdom, *The Oshawa Express* in Canada, *Pune Mirror* of India, United Press International (UPI), and *Youth Hub Africa* of Africa.

Throughout the United States, coverage of Earth Science Week programs and activities was provided by **local news organizations** such as *Alcalde* of the University of Texas; *The Argus Observer* of Oregon; *Arkansasmatters.com; Baylor Lariat* in Texas; Broomfield Veterans Memorial Museum in Colorado; *The Buckeye Lake Beacon* of Ohio; Bureau of Economic Geology at the University of Texas at Austin; Click2Houston in Texas; *Daily Camera* in Boulder, Colorado; *The Daily Nebraskan* of Nebraska; *Daily News-Miner* of Alaska; *The Daily Sentinel* of Grand Junction, Colorado; *The Daily Texan* of the University of Texas at Austin; *The Daily Times* of Tennessee; *The Denver Post* of Colorado; *Deseret News* of Utah; *Do Awesome Stuff in Austin (Do512)* of Austin, Texas; *Dover Post* in Delaware; *Edhat* in Santa Barbara, California; *El Paso Herald-Post* in Texas; *Event* of Fort Lauderdale, Florida, and of Waco, Texas; *Explore Licking County* in Ohio; *Farmington Daily Times* of New Mexico; Fauquier County Public Library in Virginia; *The Florida Times-Union* in Jacksonville; *FHSU News* of Kansas; *Greeley Tribune* of Colorado; Guthrie Public Library in Oklahoma; Heard Natural Science Museum and Wildlife Sanctuary in Texas; HeraldNet of Everett, Washington; *Huntington News* of West Virginia; *Kemmerer Gazette* in Wyoming; *Knoxville News Sentinel* of Tennessee; *Lake Powell Life* of Arizona; *Leominster Champion* of Massachusetts; Licking Park District in Ohio; *Los Alamos Daily Post* of New Mexico; *Martinsville Bulletin* in Virginia; *The Marshall Parthenon* of Huntington, West Virginia; *Martha's Vineyard Times* of Massachusetts; *Missoulian* of Montana; *Moab*

SUMMARY OF ACTIVITIES

• EXTERNAL EVALUATION OF EARTH SCIENCE WEEK 2017: KEY FINDINGS

Times-Independent of Utah; *My Eastern Oregon; The News-Reporter* of Washington, Georgia; *Noozhawk* of Santa Barbara, California; Ohio Department of Natural Resources; Oregon Coast Beach Connection, *Paulding County Progress* of Ohio; PennCurrent of the University of Pennsylvania; *Penn Live* of Pennsylvania; *Pilot Tribune* of Storm Lake, Iowa; *Post Bulletin* of Rochester, Minnesota; *The Press-Enterprise* in California; *The Prospector* of the University of Texas at El Paso; *Rapid City Journal* in Nebraska; *The Record-Courier* of Nevada; *Redlands Daily Facts* of California; *Rockford Register Star* of Illinois; *The San Pedro Sun* in California; *Scottsbluff Star Herald* of Nebraska; *Southeast Missourian* of Missouri; *State Gazette* of Dyersburg, Tennessee; *St. George News* of Utah; *Tallahassee Democrat* of Florida; *Times Union* of Albany, New York; *The Town Crier* in Manila, Arkansas; *UC Davis Library* of California; University of California Press Blog; University of Colorado at Boulder; University of Tennessee at Knoxville; *UT Daily Beacon* of Tennessee; *The Vanguard* of the University of South Alabama; *Vineyard Gazette* of Martha's Vineyard, Massachusetts; *Waco Tribune-Herald* of Texas; *Walla Walla Union Bulletin* of Washington; *Watertown Daily Times* of New York; *Winona Daily News* in Minnesota; *Wylie News* of Texas; *Wyoming Business Report* of Wyoming; and the Wyoming State Geological Survey.

Earth Science Week also was covered by **television stations** across the country, including ABC 10 News in Michigan, The Denver Channel ABC 7, KCRG TV9 ABC News of Iowa, KFYR TV of North Dakota, KREM 2 CBS of Washington, KTV News Channel 21 in Oregon, WBIR 10 News NBC in Tennessee, and WITN News of North Carolina. In addition, the event was covered by **radio stations** such as KIWA Radio of Iowa, National Public Radio, Radio 7 AM/FM in Tennessee, WGNS Radio of Tennessee, and WYSH Radio of Tennessee.

AGI distributed **press releases** to hundreds of newspapers, magazines, and other print media outlets. AGI staff also wrote **articles** for the Society of Exploration Geophysicists' *The Leading Edge,* the American Association of Petroleum Geologists' *AAPG Explorer,* and the National Earth Science Teacher Association's *The Earth Scientist* magazine. Program Manager Geoff Camphire gave interviews for publications such as *Fairfax Times* and *University of Tennessee Daily Beacon.* The articles highlighted Earth Science Week activities and the program theme.

More than 43,500 copies of the **Earth Science Week 2017 poster,** featuring a geoscience learning activity in addition to promotional content, were distributed as inserts in publications carrying articles about the event, such as NESTA's *The Earth Scientist, GSA Today, AAPG Explorer,* and AGI's *Earth* magazine. In addition, the poster image appeared in the Society of Exploration Geoscientists' *The Leading Edge,* which reaches some 100,000 people in print and online.

External Evaluation of Earth Science Week 2017: Key Findings

Following the event, AGI secured an independent contractor, PS International, to complete a formal external evaluation of Earth Science Week 2017, as it has in past years. Participants were invited to participate in a survey in the closing months of 2017, with a valid response rate of 6.5 percent.

Results were overwhelmingly positive. Participation remained strong. Comparing participation last year and plans for next year, 81 percent of survey respondents said they anticipate either increasing or maintaining level participation.

A large majority (89 percent) said Earth Science Week offers opportunities for teaching and promoting Earth science that they would not have otherwise. Similarly, 92 percent said program resources and activities are very or somewhat important to educating students about geoscience.

Eighty-six percent of respondents rated the program's overall usefulness as "excellent" or "good." When respondents rated nearly 20 key items from the Earth Science Week Toolkit and Website — such as posters, disks, and online activities — all were rated "very useful" or "useful" by strong majorities of participants.

Participants said they were active during Earth Science Week. Many reported specific activities that were highly active. For example, 95 percent reported activities categorized as "most active" (e.g., field trips and outside lessons), "active" (e.g., external speakers and open house discussions), or "somewhat active" (e.g., lesson plans and kit distribution).

Asked how Earth Science Week might be improved, respondents advocated additional program materials and activities, as well as increasing communication and promotion. AGI uses these evaluation findings to fuel continual improvement of the program.

SUMMARY OF ACTIVITIES
• EARTH SCIENCE WEEK SPONSORS

Earth Science Week Sponsors

United States Geological Survey

National Aeronautics and Space Administration

National Park Service

American Association of Petroleum Geologists Foundation

American Geophysical Union

Society for Mining, Metallurgy, and Exploration

Howard Hughes Medical Institute

Association of American State Geologists

Geological Society of America

Archaeological Association of America

AmericaView

ExxonMobil

Earth Science Week Program Partners

American Association of Petroleum Geologists Foundation

American Geophysical Union

American Geosciences Institute

American Institute of Professional Geologists

AmericaView

Archaeological Institute of America

Association of American Geographers

Association of American State Geologists

CLEAN Network

Consumer Energy Alliance

Critical Zones Observatories

EarthScope

Flyover Country

Geological Society of America

Google

Howard Hughes Medical Institute

Incorporated Research Institutions for Seismology

Minerals Education Coalition

National Cave and Karst Research Institute

National Earth Science Teachers Association

National Energy Education Development Project

National Public Radio's Science Friday

National Science Teachers Association

National Wildlife Federation

Society for Mining, Metallurgy, and Exploration

Society of Exploration Geophysicists

Society of Petroleum Engineers

Soil Science Society of America

Switch Energy Alliance

UNAVCO

U.S. Bureau of Land Management

U.S. Department of Energy

U.S. Geological Survey

U.S. National Aeronautics and Space Administration

U.S. National Oceanographic and Atmospheric Administration

U.S. National Park Service

U.S. Natural Resources Conservation Service

SUMMARY OF ACTIVITIES
• EARTH SCIENCE WEEK 2017 EVENTS AND ACTIVITIES BY STATE AND TERRITORY

Earth Science Week 2017 Events and Activities by State and Territory

While it is impossible to track all Earth Science Week activities in the United States, major activities across the country included:

Alabama
- The Anniston Museum held an event to celebrate paleontology and the importance of fossils in honor of National Fossil Day. Activities included identifying fossil types, creating your own fossil, and visiting with "living fossils."
- The Alabama Museum of Natural History hosted a celebration of National Fossil Day at the University of Alabama.
- The University of South Alabama Archaeology Museum welcomed visitors of all ages to participate in International Archaeology Day on October 21. The event offered a variety of indoor and outdoor hands-on activities for visitors to try.
- The Geological Survey of Alabama distributed Earth Science Week Toolkits to science educators.
- Earth Science Week 2017 Toolkits were distributed to students, teachers, and others by representatives of the International Association of Hydrogeologists, U.S. Chapter.
- The University of Alabama's Smith Hall Museum of Natural History hosted a National Fossil Day event celebrating fossil appreciation and stewardship. The public was invited to meet a paleontologist and view fossils.

Alaska
- The governor of Alaska issued a perpetual proclamation of Earth Science Week.
- Just in time for International Archaeology Day, the University of Alaska Museum highlighted archaeology in Alaska.
- The Department of Anthropology at the University of Alaska Anchorage sponsored an event where visitors could find artifacts in a mock dig, learn about the human skeleton, how to make a stone arrowhead, what animal bones can tell you, how to make a mask, and more.
- The Alaska Division of Geological & Geophysical Surveys distributed Earth Science Week Toolkits to science educators.

Arizona
- The Museum of Northern Arizona held a "Fossil Day" with exciting kids' programs, hands-on activities, and creative crafts.
- M.J. Tykoski, 2017 Winner of the American Geosciences Institute Edward C. Roy, Jr. Award For Excellence in K-8 Earth Science Teaching, wrote about the importance of earth and space sciences in education in her blog "Earth and Space Sciences: The Forgotten STEM Subjects."
- Grand Canyon National Park hosted its annual Earth Science Week and National Fossil Day celebration. Activities included ranger-led hikes exploring geologic time and paleontology lectures.
- The Pueblo Grande Museum in Phoenix hosted an International Archeology Day event in collaboration with the Central Arizona Society of the Archaeological Institute of America that featured archaeological demonstrations, children's activities, tours, and more.
- Earth Science Week 2017 Toolkits were distributed to students, teachers, and others by Carla McAuliffe of the National Association of Earth Science Teachers.

Arkansas
- The Arkansas Geological Survey promoted Earth Science Week among teachers and citizens with the distribution of related instructional materials.

California
- The Dr. John D. Cooper Archaeological and Paleontological Center celebrated International Archaeology Day and National Fossil Day with a free educational family event.
- San Bernardino County Museum participated in Earth Sciences Week by hosting activities throughout the week centered around becoming stewards of the Earth. Programs throughout the week challenged guests on geology and paleontology fun facts through hands-on and hands-off activities.
- Shields Library celebrated Geologic Map Day by displaying various geologic maps that show water-bearing formations, seismic hazard evaluations, coastal erosion, earthquakes, and more.
- The Santa Barbara Museum of Natural History, hosted a lineup of activities planned in celebration of National Fossil Day. Some museum fan favorites included Bug Boot Camp; a Science on Site event; and Eyes in the Sky, where guests can get an up-close look at owls, hawks and other local birds of prey.
- The California Geological Survey distributed Earth Science Week Toolkits to science educators.
- The International Archaeology Day & Heritage Month Expo held an archaeology fair in Riverside.

SUMMARY OF ACTIVITIES
• EARTH SCIENCE WEEK 2017 EVENTS AND ACTIVITIES BY STATE AND TERRITORY

This event included 28 exhibitors from academic, government, and private organizations. Attendees had the chance to throw atlatl darts, decipher ancient writings, interpret archaeological artifacts and write like a Roman centurion among other activities.
- Roxie Khalili of Foster City, California, won first place in the Earth Science Week photo contest with her image of human-built infrastructure surrounding the lagoon and its inhabitants.

Colorado
- Aiming to get students interested in STEM careers, Consumer Energy Alliance and Consumer Energy Foundation partnered with Energy 360 and the University of Colorado's Global Energy Management Program to host a free Energy Day Festival, featuring more than 54 interactive demonstrations and exhibits, at East High School.
- The Denver Museum of Nature and Science announced the name of a fossilized triceratops found buried in Thornton to coincide with National Fossil Day.
- For National Fossil Day, Colorado Canyons Association Educational Director Rob Gay led a hike on the Trail Through Time in Rabbit Valley.
- Museums of Western Colorado, Colorado Canyons Association and the John McConnell Math & Science Center sponsored a series of educational events in Mesa County in honor of Earth Science Week.
- In honor of National Fossil Day, the University of Colorado Museum of Natural History held an educational event open to the public.
- The Geological Society of America encouraged geocachers around the world to participate in International EarthCache Day during Earth Science Week.
- Museums of the West held events featuring free educational activities for families across Colorado.
- Earth Science Week 2017 Toolkits were distributed to students, teachers, and others by representatives of the American Institute of Professional Geologists, the Geological Society of America, the Society of Mineral Museum Professionals, and the Society for Mining, Metallurgy & Exploration.
- U.S. Rep. Jared Polis (D-Colorado) submitted H. Res. 556, on behalf of himself and Reps. Barbara Comstock (R-Virginia) and Dan Lipinski (D-Illinois), advancing a resolution expressing support for designation of the week of October 8-14, 2017, as Earth Science Week.

Connecticut
- The Westport Library sponsored a local celebration of International Archaeology Day featuring lectures, demonstrations and interactive experiences throughout the library.
- The Talcottville Congregational Church hosted a three-mile archaeological hike to see the Talcottville gorge, historic bridges and buildings, and the sites of former mills.
- The Connecticut Geological Survey promoted Earth Science Week and distributed kits among educators in the state.

Delaware
- Fort DuPont State Park partnered with the Delaware State Parks Time Travelers program to celebrate International Archaeology Day at the Fort Dupont State Park Grassdale Center.
- The governor of Delaware issued a perpetual proclamation of Earth Science Week.
- In addition to promoting Earth Science Week by distributing information and materials, the Delaware Geological Survey recommended fossil hunters to visit the Chesapeake and Delaware Canal in celebration of National Fossil Day.

District of Columbia
- The National Park Service and AGI hosted exhibits and activities at the National Fossil Day celebration on the National Mall.
- Numerous federal agencies, including the U.S. Geological Survey, NASA, and the National Park Service, supported Earth Science Week with the provision of teaching materials, distribution of kits, and events and activities.
- Earth Science Week 2017 Toolkits were distributed to students, teachers, and others by D.C. Public Schools and the American Geophysical Union.

Florida
- The Florida Department of Environmental Protection's Florida Geological Survey hosted an Open House celebrating Earth Science Week. Informal talks, interactive exhibits, a drilling demonstration and computer models of geoscience in Florida, and a fossil dig were among several pastimes at the event to showcase activities and tools researchers use to learn about the earth, its systems and its processes at its museum. Guests were also invited to bring their own rock, fossil or mineral for a geologist to identify.
- The Museum of Discovery and Science hosted events for Earth Science Week on Saturday, October 14 and Sunday, October 15.
- About 1,120 people attended the Florida Museum of Natural History's weekend-long National Fossil Day celebration.
- The Archaeological Institute of America Jacksonville Chapter presented International Archaeology Day at the Beaches Museum & History Park. Activities included mock digs and pottery making,

SUMMARY OF ACTIVITIES

• EARTH SCIENCE WEEK 2017 EVENTS AND ACTIVITIES BY STATE AND TERRITORY

artifact identification, as well as a lecture in the Beaches Museum Chapel.
- The Florida Geological Survey promoted Earth Science Week by distributing educational materials.

Georgia
- Earth Science Week 2017 Toolkits were distributed to students, teachers, and others by representatives of AASP-The Palynological Society and the National Association of Black Geoscientists.
- Emory University hosted "Dig It!" their annual Archaeology Festival featuring presentations, a "highlights" tour of the Carlos Museum Mediterranean collection, as well as a Mediterranean-themed lunch.
- Students at Wilkes Primary School in Washington held a No Child Left Inside Day event sponsored by the Iris Garden Club, featuring outdoor learning activities including soil classifications, tree ring observations, and leaf rubbings guided by soil scientists, foresters, geologists, and others.
- The Anthropology Program and Student Anthropology Club at Kennesaw State University hosted an archaeology booth at the Alpharetta Farmers Market to celebrate International Archaeology Day.
- The staff at Fort Frederica National Monument, a unit of the National Park Service, arranged an event "Archaeology in Georgia: Past and Present," on Saturday, October 21. There were activities for children, exhibits, and presentations by area archaeologists and historian.

Hawaii
- The Hawaii State Historic Preservation Division welcomed the public to come and hear about how they register archaeological sites, manage GIS data and conduct historic preservation reviews.

Idaho
- Idaho Museum of Natural History celebrated International Archaeology Day by sponsoring a day of exploring Ancient Egypt and the methods of mummification. Those who attended this family friendly event brought their own doll, of any kind, to make into a fun Halloween decoration.
- The Idaho Geological Survey distributed Earth Science Week educational materials.
- Hagerman Fossil Beds National Monument celebrated National Fossil Day with the Hagerman Valley Foundation's Fossil Daze. The Event included a parade through town, as well as an afternoon activities and booths in the Hagerman City Park.

Illinois
- To celebrate this year's International Archaeology Day, Monmouth College hosted a talk on the "Archaeology of the Stars" exploring how ancient Greek and Roman civilizations interacted with the night sky.
- Burpee Museum of Natural History hosted a National Fossil Day celebration. The event included local fossil collectors and exhibitors, hands-on fossil activities, the chance to have your fossils identified, and Jane's real skull on display.
- U.S. Rep. Jared Polis (D-Colorado) submitted H. Res. 556, on behalf of himself and Reps. Barbara Comstock (R-Virginia) and Dan Lipinski (D-Illinois), advancing a resolution expressing support for designation of the week of October 8-14, 2017, as Earth Science Week.
- The Illinois Field Museum invited the public to their Meet a Scientist event to see local Ice Age fossils from Northern Illinois and Indiana as well as Paleozoic vertebrates from the famous Mazon Creek.
- The governor of Illinois issued a perpetual proclamation of Earth Science Week.

Indiana
- Indiana hosted the 2017 Midwest Archaeological Conference at the The Alexander Hotel.
- A local celebration of National Fossil Day was held at Falls of the Ohio State Park.
- The Children's Museum of Indianapolis celebrated International Archaeology Day with fun and educational archaeology activities.

Iowa
- O'Brien County Conservation Board celebrated International Archaeology Day at the Prairie Heritage Center by engaging participants in learning about and playing a wide variety of games.
- Iowa's National Park Service declared October 11 National Fossil Day.
- The University of Iowa Paleontology Repository hosted a ton of fossils brought in by fossil hunters of all ages from across the country.
- The University of Northern Iowa and BMC Aggregates partnered for the "Sunday at the Quarry" event in Waterloo to offer the public opportunities to dig for fossils, search for geodes, and interact with Earth systems.

Kansas
- The Kansas Geological Survey obtained and distributed Earth Science Week Toolkits for educators.
- The University of Kansas Natural History Museum celebrated National Fossil Day with a number of activities for the public including fossil casting and an excavation of fossilized shark teeth.

Kentucky
- The Kentucky Geological Survey distributed educational materials and hosted its annual Open House to celebrate Earth Science

SUMMARY OF ACTIVITIES
EARTH SCIENCE WEEK 2017 EVENTS AND ACTIVITIES BY STATE AND TERRITORY

Week at the Mining and Mineral Resources Building. Attendees were invited to browse the rock and fossil collections.

Louisiana
- The Louisiana Geological Survey distributed Earth Science Week Toolkits to schools across the state.
- Poverty Point World Heritage Site in Pioneer hosted activities for International Archeology Day, including a guided tram tour with the station archaeologist, a demonstration of prehistoric tools, and more.

Maine
- Maine State Museum was the site for Maine Earth Science Day 2017, a day of exhibits and hands-on activities for all ages on topics ranging from minerals and gems to maps to solar electricity.
- The Maine Geological Survey promoted Earth Science Week among citizens across the state.

Maryland
- The governor of Maryland, Lawrence Hogan, declared October 8-14 to be Earth Science Week.
- Maryland National Capital Park and Planning Commission Dinosaur Park participated in the National Fossil Day celebration at the 2017 National Mall event.
- The Maryland Geological Survey promoted Earth Science Week among citizens across the state.
- Maryland Science Center hosted Dino Day at the Center for National Fossil Day.
- The Natural History Society of Maryland hosted a National Fossil Day event with Paleontologist Dr. Stephen J. Godfrey.
- Earth Science Week 2017 Toolkits were distributed to students, teachers, and others by representatives of the Society of Vertebrate Paleontology.

Massachusetts
- The Oak Bluffs Library held a National Fossil Day event for Earth Science Week. People were invited to and share their fossils with collectors and listen to presentations.
- Harvard Museum of Natural History promoted National Fossil Day with a day of ancient trilobites, sea scorpions, saber-toothed cats, giant ground sloths, and more.

Michigan
- The Marquette Regional History Center hosted a local celebration of International Archaeology Day with a day of hands–on activities, projects, booths, and actual artifacts.
- The Geology Department at Grand Valley State University sponsored a lecture series, featuring alumni presenters, in recognition of Earth Science Week .
- The Michigan Geological Survey promoted Earth Science Week and distributed kits among educators statewide.
- Michigan Technological University hosted an Energy Day event in September as part of its CareerFEST. Students had opportunities to learn about the energy industry and different job and career opportunities.

Minnesota
- The Minnesota Geological Survey promoted Earth Science Week by distributing kits to teachers in the state.
- Winona State University sent out a press release on a 66 million-year-old fossil that was to be displayed in the atrium of the Science Lab Center for National Fossil Day.
- Winona State University's National Fossil Day exhibit drew a steady crowd of students, faculty and interested community members.
- Education Minnesota promoted the Earth Science Week photography contest to teachers and students in the state.
- Earth Science Week 2017 Toolkits were distributed to students, teachers, and others by representatives of the National Association of Geoscience Teachers.

Mississippi
- The Bollinger County Museum of Natural History celebrated International Archaeology Day with a presentation by Mike Comer, "Mississippi Mound Building Indians." Hands-on children's activities, and other puzzles, games, and crafts appropriate to the archaeology theme were also ongoing throughout the day.

Missouri
- The governor of Missouri declared October 8-14 to be Earth Science Week.
- The Bollinger County Museum of Natural History held a National Fossil Day celebration in Marble Hill, Missouri on Saturday, October 14 from 12:00 (noon) to 4:00 p.m.
- The Missouri Geological Survey promoted Earth Science Week among teachers and citizens with the distribution of educational materials.
- The University of Missouri Museum of Art and Archaeology celebrated International Archaeology Day, disseminating information on archaeology and related activities statewide.

Montana
- The University of Montana Paleontology Center hosted a celebration of National Fossil Day at the Charles H. Clapp Building to educate the public about fossils. At the free event, activities included fossil identifications, tours of displays and research collections, as well as a chance to hear about research going on at the center.

SUMMARY OF ACTIVITIES

EARTH SCIENCE WEEK 2017 EVENTS AND ACTIVITIES BY STATE AND TERRITORY

- Montana State University's Museum of the Rockies invited children to earn "Junior Paleontologist" badges by participating in National Fossil Day activities.

Nebraska
- The University of Nebraska State Museum celebrated National Fossil Day on Oct. 5, with activities for all ages. The museum showed kids how to dig for fossils, how to identify a fossil, as well as allowed people to bring in fossils to learn more about them.
- Agate Fossil Beds National Monument celebrated National Fossil Day with special events and programs.

Nevada
- The governor of Nevada, Brian Sandoval, issued a perpetual proclamation of Earth Science Week.
- The Nevada Bureau of Mines and Geology sponsored an educational Earth Science Week field trip to Carson Valley, for people of all ages.
- Tule Springs Fossil Beds National Monument celebrated National Fossil Day at the Las Vegas Museum of Natural History. The public had opportunities to observe paleontologists doing fossil prep, explore hands-on fossil carts, and witness a shark feeding demonstration.

New Hampshire
- The New Hampshire Geological Survey promoted awareness of Earth Science Week among residents and community members.

New Jersey
- During Earth Science Week, the New Jersey State Museum had staff and volunteers stationed at their public laboratory from Tuesday, October 10, through Friday, October 13.
- The New Jersey State Museum invited the public to join them in celebrating International Archaeology Day 2017 for activities including a pottery reconstruction interactive, a scavenger hunt focused on their archaeology exhibition, a film, gallery tours of the archaeology exhibition, artifact identification and an interactive interpreting artifacts "recovered" from an archaeological site.
- The New Jersey Geological and Water Survey distributed Earth Science Week Toolkits and Geologic Map Day materials to teachers.

New Mexico
- International Archaeology Day was celebrated at Aztec Ruins National Monument with activities including an East Ruins tour and a flint knapping demonstration.
- Mesalands Community College Dinosaur Museum invited visitors attend their National Fossil Day event and see genuine fossils collected from students enrolled in their Paleontology program as wells as life-sized bronzes created through the Fine/Arts Bronze program at Mesalands Community College.

New York
- In celebration of International Archaeology Day, the Underground Railroad History Project of the Capital Region, Inc. invited the public to an interactive exhibit exploring 19th century life through historical artifacts found during excavations conducted by the University at Albany.
- On International Archaeology Day, the Depauville Free Public Library hosted discussions by Kenneth J. Knapp on the archaeology of northwest Jefferson County.
- The American Museum of Natural History held a National Fossil Day-inspired event to help students understand how studying fossils teaches us about the history of life, past climates, and ancient landscapes. The day consisted of activities such as the exploration of fossil formation, extinction theories, fossil diversity and much more.
- The Museum of the Earth and the Paleontological Research Institution hosted Fossil Mania, a three-hour event that educated museum visitors about fossil types, & formation with hands-on specimen-driven activities.
- Earth Science Week 2017 Toolkits were distributed to students, teachers, and others by representatives of the Paleontological Research Institution.

North Carolina
- The Aurora Fossil Museum celebrated National Fossil Day with a variety of special guests, live music, events and activities planned for families and kids. The free event is annually through a partnership with the National Park Service.
- Imagination Station Science and History Museum hosted a National Fossil Day inspired Educator Trek – Fossil Hunt through the waters of Green Mill Run.
- Town Creek Indian Mound State Historic Site, North Carolina Department of Natural & Cultural Resources, Historic Sites & Properties sponsored International Archaeology Day at Town Creek.
- The North Carolina Geological Survey promoted awareness of Earth Science Week among residents and community members.
- Earth Science Week 2017 Toolkits were distributed to students, teachers, and others by representatives of the History of Earth Sciences Society.

North Dakota
- North Dakota's governor issued a perpetual proclamation of Earth Science Week.

SUMMARY OF ACTIVITIES
EARTH SCIENCE WEEK 2017 EVENTS AND ACTIVITIES BY STATE AND TERRITORY

- In honor of National Fossil Day, the North Dakota Geological Survey (NDGS) spent time talking to community members and kids at the Heritage Center, educating them about fossils. Several paleontologists spent time examining rocks or what people thought to be fossils that they had found.
- The Main Library hosted Paleozoic Life on Land & Sea — Kids, a National Fossil Day celebration, on Saturday, October, 7 with Dr. Lydia Tackett and Jessie Rock from North Dakota State University's Department of Geosciences.

Ohio
- The Ohio Department of Natural Resources (ODNR) Division of Geological Survey distributed Earth Science Week Toolkits to schools and others.
- The Ohio Division of Geological Survey celebrated Earth Science Week by leading the public on a number of geologic hikes and fossil collection throughout the state at parks, preserves and other important geologic sites as well as hosting talks and other educational events throughout the state.
- As part of Earth Science Week, Survey geologists Andy Nash and Nate Erber led explorers on a hike to examine the natural cliffs of shale and glacial till carved out by Lobdell Run Creek and discuss how that landscape changed over time due to glaciers, erosion, and human activity. Time was also spent in the stream getting up close and personal with the exposed cliff faces.

Oklahoma
- Oklahoma's governor issued a perpetual proclamation of Earth Science Week.
- The Guthrie Public Library celebrated Earth Science Week with a Clouds Above Program.
- The Oklahoma Geological Survey distributed Earth Science Week Toolkits to educators and students.
- The Sam Noble Museum in Norman celebrated International Archaeology Day with family-oriented activities including flint knapping and atlatl demonstrations.
- Earth Science Week 2017 Toolkits were distributed to students, teachers, and others by representatives of the Society for Sedimentary Geology and the Society of Exploration Geophysicists.

Oregon
- The governor of Oregon, Kate Brown, proclaimed October 8-14 to be Earth Science Week.
- The Cannon Beach History Center and Museum partnered with the Archaeology Institute of America for International Archaeology Day. To celebrate, the museum held a pop-up exhibit on the history of archaeology in Egypt and hosted a presentation on Dr. Sarah Sterling's work and research on pyramids in Egypt.
- Oregon's new interactive Lidar Data Viewer launched during Earth Science Week.
- The Oregon Department of Geology and Mineral Industries released an interactive geologic map of Oregon for Geologic Map Day. The application, a decade in the making, allows the public the geologic story of Oregon and access supplementary information with a click of a mouse.

Pennsylvania
- The American Research Center in Egypt, Pennsylvania Chapter (ARCE-PA) sponsored a lecture by Dr. Kevin M. Cahail titled "Of Scribes and Stables: Recent Discoveries in a New Kingdom Cemetery at South Abydos" at the University of Pennsylvania Museum of Archaeology and Anthropology in honor of International Archaeology Day.
- The Historic Preservation Trust of Berks County (HPTBC) held a public program celebrating International Archaeology Day where volunteer archaeologists presented their latest findings at the Mouns Jones House Root Cellar.
- In honor of National Fossil Day, the education department of the State Museum of Pennsylvania in Harrisburg ran various programs concentrated on ancient fishes of the Paleozoic Era.
- The Penn Museum participated in International Archaeology Day with a special event, co-sponsored by the Philadelphia society of the Archaeological Institute of America, celebrating their newest exhibition "Moundbuilders." This exhibit offers insight into the work of archaeology in North America and around the world.
- Earth Science Week Toolkits were distributed among science teachers statewide by the Pennsylvania Geological Survey.
- The governor of Pennsylvania declared October 8-14 to be Earth Science Week.

Rhode Island
- University of Rhode Island professors, Kris Bovy (Anthropology) and Bridget Buxton (History) held a talk on why archaeological investigation can be relevant to our own lives today.

South Carolina
- The South Carolina Geological Survey celebrated Earth Science Week and Geologic Map Day by distributing Earth Science Week Toolkits to educators.
- Redcliffe Plantation State Historic Site hosted "Beyond the Gravestone" in celebration of International Archaeology Day. The tour of the Beech Island Cemetery, also known as the Hammond Family

SUMMARY OF ACTIVITIES
EARTH SCIENCE WEEK 2017 EVENTS AND ACTIVITIES BY STATE AND TERRITORY

Cemetery, included a scavenger hunt and discussions of cemetery architecture and symbolism.

South Dakota
- South Dakota's governor issued a perpetual proclamation of Earth Science Week.
- In participation with National Fossil Day Student volunteers from the Paleo Club at South Dakota School of Mines and Technology set up tables to identify rocks and fossils brought in by members of the public and gave brief descriptions of the environments that created or fossilized said specimens.
- Mitchell Prehistoric Indian Village hosted an International Archaeology Day event including tours of their dig site and museum.

Tennessee
- The governor of Tennessee, Bill Haslam, declared October 8-14 to be Earth Science Week in Tennessee.
- The Tennessee Department of Environment and Conservation's Tennessee Geological Survey distributed Earth Science Week Toolkits to educators.
- On October 10, the Muse Knoxville celebrated "No Child Left Inside Day" with an outdoors day themed around pumpkins.
- The University of Tennessee's McClung Museum of Natural History and Culture with support from the Archaeological Institute of America's East Tennessee Society hosted the "Can You Dig It?" event to celebrate International Archeology Day and National Fossil Day. Participants were invited to bring in rocks, fossils or other artifacts for identification by archaeologists and paleontologists and their graduate students.
- In conjunction with the "Can You Dig It?" event, University of Tennessee, Knoxville graduate students and professors hosted a STEAM panel discussion with women scientists.
- The Muse Knoxville, a children's science museum, celebrated No Child Left Inside Day with an outdoors day themed around pumpkins.

Texas
- In partnership with the Austin Earth Science Week Consortium, the Bureau of Economic Geology held a career event for middle school students. Earth science professionals gave presentations about their careers, and students were able to participate in hands-on activities through various exhibits. Students heard from local STEM (science, technology, engineering, and math) professionals including geologists, geophysicists, engineers, hydrologists, meteorologists, paleontologists, water conservation specialists, biologists, and aerospace engineers.
- Baylor University's department of geosciences hosted a series of activities and events to celebrate Earth Science Week throughout the week at the Mayborn Museum. Visitors were invited to engage with geoscience professors and students and participate in hands-on activities.
- The Dallas Paleontological Society hosted a National Fossil Day celebration at the Heard Natural Science Museum & Wildlife Sanctuary.
- The Texas Memorial Museum held a public event, in celebration of National Fossil Day, with activities including fossil identifications, story time, dig pit and gallery talks.
- Locals, students, visiting scientists, fossil hunters and enthusiasts gathered in Fannin County for a week of activities and events to partake in Earth Science Week.
- Organizations, including the Bertha Voyer Memorial Library, the Blackland Prairie Chapter of Texas Master Naturalist, and the Ladonia Volunteer Fire Department, contributed to the 2017 celebration of Earth Science Week.
- The El Paso Museum of Archaeology celebrated at the museum on October 21 for Texas Archaeology Month and International Archaeology Day.
- The Houston Geological Society held various events for Earth Science Week with partners in the Houston area, including field trips and a special celebration at the Houston Museum of Natural Science.
- The Energy Day Festival, sponsored by more than 100 businesses in Houston, brought thousands of people out to view demonstrations, learn about energy science, and play games to compete for prizes and giveaways.
- The Just Energy Foundation awarded top-performing students of the Science and Engineering Fair of Houston during the Energy Day Festival.
- Waco Mammoth National Monument invited the public to celebrate Geologic Map Day by completing a mapping activity.
- Texas Memorial Museum celebrated National Fossil Day in Austin, where the public was invited to meet paleontologists, learn about the prehistoric whorl-toothed sharks of Texas and other fossil finds.
- The Geological Sciences department at University of Texas, El Paso celebrated Earth Science Day showcasing the department and teaching the community about the importance of Earth Science. Multiple events were held throughout a four-hour period, and took place inside the Geology building and outside on the lawn in front of the building.
- Earth Science Week 2017 Toolkits were distributed to students, teachers, and others by representatives of the Petroleum History Institute.

SUMMARY OF ACTIVITIES
• EARTH SCIENCE WEEK 2017 EVENTS AND ACTIVITIES BY STATE AND TERRITORY

- Students at Cavazos Elementary School in Nolanville participated in No Child Left Inside Day by taking part in a number of engaging science activities around their school grounds.

Utah
- In celebration of National Public Lands Day and in preparation for National Fossil Day, The Bureau of Land Management- Moab Field Office hosted a community service event to reroute a trail at the Mill Canyon Dinosaur Tracksite.
- The Bureau of Land Management- Moab partnered with the Museum of Moab and Utah Friends of Paleontology to offer free, guided tours of the Mill Canyon Dinosaur Tracksite to celebrate National Fossil Day.
- Glen Canyon National Recreation Area hosted a celebration of National Fossil Day at the Carl Hayden Visitor Center where kids could earn their Junior Ranger Paleontology Badge, dig for fossils and decorate a Jurassic landscape.
- Capitol Reef National Park celebrated Earth Science Week with a variety of activities. Visitors to the park were invited to discover how the park was formed during geology programs each day.
- On October 11th, Capitol Reef National Park Rangers led a 90-minute hike on the Fremont River Trail. Participants learned about the geology and fossils that tell of the park's ancient environments.
- Capitol Reef National Park Rangers hosted an hour-long program on the fossils found in the park on October 12th in the campground amphitheater.
- The Utah Geological Society distributed Earth Science Week Toolkits to educators and students.

Vermont
- The Vermont Geological Survey shared geoscience materials to celebrate Earth Science Week.

Virginia
- In celebration of International Archaeology Day on Saturday, October 14, the Virginia Museum of Natural History held a fun afternoon event for visitors to learn about the museum's fascinating archaeology research and collections while getting a glimpse at what goes on behind-the-scenes in the labs and collections areas.
- To celebrate Earth Science Week and Geologic Map Day, the USGS, in collaboration with the American Geosciences Institute, hosted a "Geologic Open House" on Thursday, October 12, 2017 at Great Falls Park, near McLean, Va.
- Torpedo Factory Art Center, in partnership with the American Geosciences Institute, invited the public to attend an exciting night of music, interactive art, artist presentations, and more.
- David de Costa, of Alexandria, Virginia, won first place in the Earth Science Week visual arts contest with a creative and colorful drawing of earth, water, air, and living things.
- Tracy Peucker of Virginia Beach, Virginia, won first place in the essay contest with a paper on "The Effects of Geosciences on Landslide Prevention."
- U.S. Rep. Jared Polis (D-Colorado) submitted H. Res. 556, on behalf of himself and Reps. Barbara Comstock (R-Virginia) and Dan Lipinski (D-Illinois), advancing a resolution expressing support for designation of the week of October 8-14, 2017, as Earth Science Week.
- James Monroe's Highland held an Archaeology Open House to celebrate Virginia Archaeology Month as well as International Archaeology Day. This event featured on-site informal discussions of current archaeological research at James Monroe's Highland and brief site tours by the researchers.
- The Virginia Department of Mines, Minerals and Energy distributed Earth Science Week Toolkits to teachers in the state.
- Earth Science Week 2017 Toolkits were distributed to students, teachers, and others by the Society for Organic Petrology.

Washington
- The University of Puget Sound Department of Classics, in conjunction with the Center for Intercultural and Civic Engagement and the Office of Diversity and Inclusion, organized a symposium on the practices of indigenous archaeology and broader decolonizing perspectives in education for International Archaeology Day. This symposium was organized in honor of both International Archaeology Day on October 15th and Indigenous Peoples Day on October 10th.
- Gonzaga University hosted an event to celebrate the fossils found in the Inland Northwest in honor of National Fossil Day.
- Earth Science Week Toolkits were distributed among science teachers statewide by the Washington State Department of Natural Resources.

West Virginia
- Marshall University held an Earth Science Bowl on Tuesday, October 17 as part of the celebration for Earth Science Week. The contest, sponsored by the Marshall College of Science and department of geology, invited high school students, geology major college students, as well as non-majors to participate in the fun-filled event. Prize money and trophies were awarded to the first and second place teams.
- Grave Creek Mound Archaeological Complex hosted an International Archaeology Day celebration on Saturday, October 21.

SUMMARY OF ACTIVITIES
• INTERNATIONAL EVENTS

Wisconsin
- The Archaeological Institute of America, Milwaukee Society; University of Wisconsin-Milwaukee Department of Anthropology; University of Wisconsin-Milwaukee Department of Art History; and University of Wisconsin-Milwaukee FLL/Classics Program partnered to host "Down Home Archaeology: Digging into the Past with Local Archaeologists." From experimental archaeology to helping identify and analyze ancient artifacts, this event provided many fun and interactive ways learn about how local archaeologists do their research.
- Earth Science Week 2017 Toolkits were distributed to students, teachers, and others by representatives of the Soil Science Society of America.

Wyoming
- The Wyoming State Geological Survey disseminated Earth Science Week 2017 Toolkits among students and educators statewide.
- Tate Geological Museum's annual National Fossil Day open house included museum tours, face and body art, fossil hunting, treats and more.
- The Wyoming State Geological Survey and University of Wyoming Geological Museum celebrated with a geoscience event on October 14.
- Fossil Butte National Monument staff and volunteers held a celebration of National Fossil Day with various activities, including hands-on demonstrations, arts and crafts and exhibit tours of the monument museum.

International Events

According to Google Analytics, the Earth Science Week website was accessed by users in 213 countries, territories, and regions worldwide in 2017. Additional activity included:

Africa
- *Youth Hub Africa* announced the Earth Science Week 2017 Photo Contest and urged readers to participate.

Australia
- Geoscience Australia's Earth Science Week 2017 celebrated the theme of "Earth and Human Activity."
- The National Dinosaur Museum held a special midday tour focusing on Australia in celebration of National Fossil Day.
- The Australian Archaeology Association Annual Meeting 2017 was hosted by La Trobe University, coinciding with its 50th Anniversary. The conference theme was "Island to Inland: Connections Across Land and Sea."
- The Geological Survey of New South Wales hosted an Open Day in honor of Earth Science Week.

Belize
- The Belize Institute of Archaeology held their 2017 educational events during the week of October 16 through 20, International Archaeology Day concluding with a two-day event in San Pedro. Hosted by the non-profit organization Marco Gonzalez Maya Site Ambergris Caye, Ltd (MGMSAC), the event was held at El Patio Restaurant in downtown San Pedro.
- The Belize Institute of Archaeology in Orange Walk and San Pedro sponsored a tour with a lecture series to celebrate International Archaeology Day.

Brazil
- The 3rd Olympiad of Geography and the 1st Olympiad of Earth Sciences were held in the capital city of Brasília, engaging 33 teams from across Brazil.
- A lecture titled "What Is Geology?" was presented to students at the Nossa Senhora da Piedade Institute, Rio de Janeiro.
- The Prioridade Hum School in Rio de Janeiro hosted a "Geology and the Role of the Geologist in Society" discussion.
- Students of Rio de Janeiro's Educational Society listened to a presentation about geosciences and viewed rock samples. In addition, students of Bahiense High School heard a lecture on "Geologists' Work."
- Related to National Fossil Day, Brazil won "Fossil of the Day" at the 2017 United Nations Climate Change Conference.

Canada
- The Oshawa Museum in Ontario partnered with Trent University Durham and Scugog Shores Museums to host an International Archeology Day event featuring interactive displays, engaging activities, and lectures on archeology.
- The public was invited to experience archaeology hands on during International Archaeology Day at the Museum of Ontario Archaeology. Some activities included were flint knapping, archaeology demonstrations, a guided tour of the Permanent Gallery and Medway Heritage Forest, and cookie excavations.

SUMMARY OF ACTIVITIES
• INTERNATIONAL EVENTS

- BCTV News and *The Oshawa Express* promoted celebrations of International Archaeology Day, including exhibits, demonstrations, and interactive sessions.

Colombia
- The Colombian Geological Society Student Chapter at the University of Pamplona hosted Earth Science Week in Cúcuta. The program focused on university staff training on geology, what geologists do, and the importance of geology in people's lives. The university hosted lectures from Colombian Geological Survey Geologists about the Colombian Geological map, geochemistry, petrography and paleoseismology in Colombia.
- Earth Science Week-related education materials (in English and Spanish) were distribution.
- Universidad Externado de Colombia sponsored a week-long celebration of International Archaeology Day (Semana de Arqueología Externadista) which included activities such as conferences, lectures, and workshops.

Czech Republic
- The Museum of Mohelnice and Museum of Šumperk partnered in hosting two days of programs for schools, families, children and adults to celebrate International Archaeology Day.
- The City of Prague Museum celebrated International Archeology Day by hosting an all-day series of workshops for people of all ages.
- Kulturní zařízení Kadaň sponsored International Archaeology Day in the KD Střelnice courtyard with a fair of activities for kids and adults. At the event, visitors could become a knight or soldier, medieval farmer, potter of all (pre)historical periods, stove maker and modern archaeologist.

Ecuador
- Archaeology Vacations held an excavation of a Manteño Period structure in Agua Blanca, Ecuador.

Georgia
- The University of Georgia's departments of Archaeology, Anthropology and Art held an educational event in the village Samshvilde. The event included various workshops, night camping, hiking on the archaeological site Samshvilde as well as a lecture.

Greece
- The Ephorate of Antiquities of Achaea, Ministry of Culture and Sports, Patras, Greece celebrated this year's International Archaeology Day with a two-session educational event that was held on October 16th and 20th, 2017 at the Mycenaean archaeological park of Voudeni, Patras.

India
- Set to coincide with Earth Science Week, the Centre for Education and Research in Geosciences (CERG) and Fergusson College jointly hosted Geo Week for the first time — a five-day event that featured exhibitions, talks and film screenings to promote the subject.

Iran
- The Khorramshahr wing of Iran's Cultural Heritage, Handicrafts and Tourism Organization in Khuzestan Province organized the scientific "Analysis on Architecture and Drainage System at Chogha Zanbil" in celebration of International Archaeology Day.

Ireland
- Earth Science Week events took place in museums, libraries, schools, universities, geoparks and more, including urban geowalks, family activity days, talks, conservation days and field trips.

Peru
- Nivín, Peru hosted a week of activities for Earth Science Week's International Archaeology Day.

Trinidad and Tobago
- AAPG and Shell partnered to provide STEM (science, technology, engineering, and math) education activities including lectures and field trips for high school students.

United Kingdom
- The Geological Society of London (GSL) announced the winners of the 2017 Earth Science Week photography competition, "Our Restless Earth." The 12 winning images represent the dynamic processes that have shaped the UK and Ireland over its tectonic history, from ancient volcanic activity to ice age glaciers. Winning photos are included in GSL's 2018 calendar.
- Sedgwick Museum of Earth Science celebrated Earth Science Week with a new activity trail exploring our ever-changing planet. Visitors were welcomed to visit the museum's new Earthquake monitor and see seismic activity around the world in real time as well as learn about plate tectonic, volcanoes, and earthquakes.
- *The Northern Times* reported on activities at Geoparks during Earth Science Week 2017.
- Earth Science Week 2017 Photo Contest entries were featured in *The Guardian* and *The Daily Mail*.
- This year's Earth Science Week saw around 60 events in the United Kingdom and Ireland, ranging from craft workshops to geowalks, exhibitions, talks and quiz nights.

www.ingramcontent.com/pod-product-compliance
Lightning Source LLC
Chambersburg PA
CBHW051830210526
45473CB00005B/1820